Animals in My Yard

Coyotes

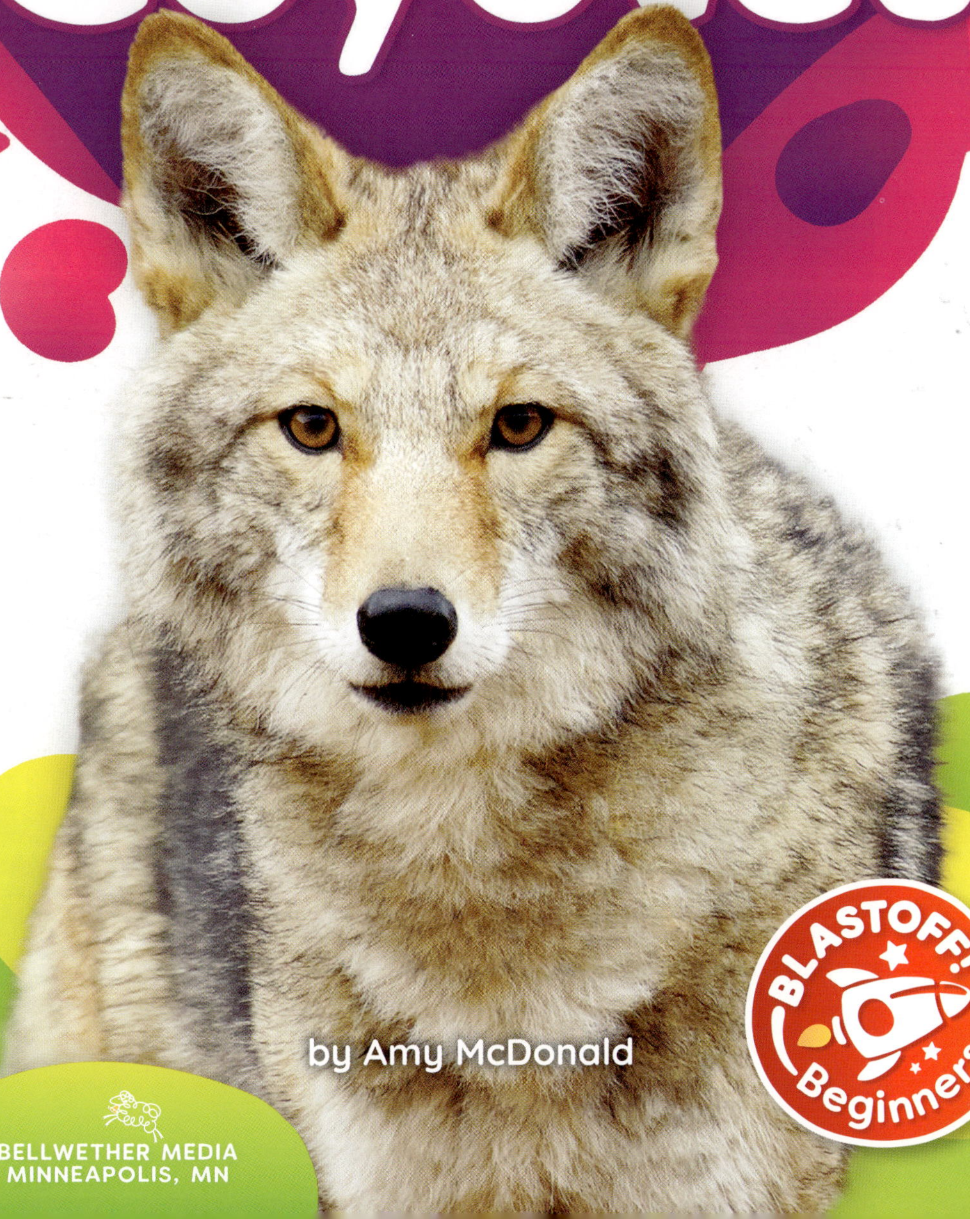

by Amy McDonald

BLASTOFF! Beginners

BELLWETHER MEDIA
MINNEAPOLIS, MN

Blastoff! Beginners are developed by literacy experts and educators to meet the needs of early readers. These engaging informational texts support young children as they begin reading about their world. Through simple language and high frequency words paired with crisp, colorful photos, Blastoff! Beginners launch young readers into the universe of independent reading.

Sight Words in This Book

a	eat	in	one	this
am	find	is	say	well
and	good	it	see	what
are	have	like	she	with
at	here	no	that	
did	I	not	they	

This edition first published in 2022 by Bellwether Media, Inc.

Library of Congress Cataloging-in-Publication Data

Names: McDonald, Amy, author.
Title: Coyotes / by Amy McDonald.
Description: Minneapolis, MN : Bellwether Media, 2022. | Series: Blastoff! beginners : Animals in my yard | Includes bibliographical references and index. | Audience: Ages 4-7 | Audience: Grades K-1
Identifiers: LCCN 2021000779 (print) | LCCN 2021000780 (ebook) | ISBN 9781644874738 (library binding) | ISBN 9781648343810 (ebook)
Subjects: LCSH: Coyote--Juvenile literature.
Classification: LCC QL737.C22 M3823 2022 (print) | LCC QL737.C22 (ebook) | DDC 599.77/25--dc23
LC record available at https://lccn.loc.gov/2021000779
LC ebook record available at https://lccn.loc.gov/2021000780

Editor: Christina Leaf   Designer: Brittany McIntosh

Printed in the United States of America, North Mankato, MN.

Table of Contents

Coyotes!

Is that a dog?
No! It is a coyote!

Body Parts

Coyotes have thick fur. They have fluffy tails.

fur

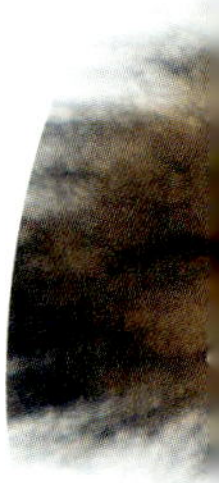

tail

Coyotes have pointy ears. They have strong noses.

ears
nose

Coyotes have
good eyes.
They see well
at night.

eyes

The Lives of Coyotes

Coyotes hunt **prey** at night. They like mice, birds, and rabbits.

prey

birds

rabbits

Coyotes are
not picky.
This one
eats trash!
Yum!

This **pup** hunts with mom. What did she find?

pup

Pups sleep
in **dens**.
They are safe!

den

Coyotes bark
and howl.
This one says,
"Here I am!"

Coyote Facts

Coyote Body Parts

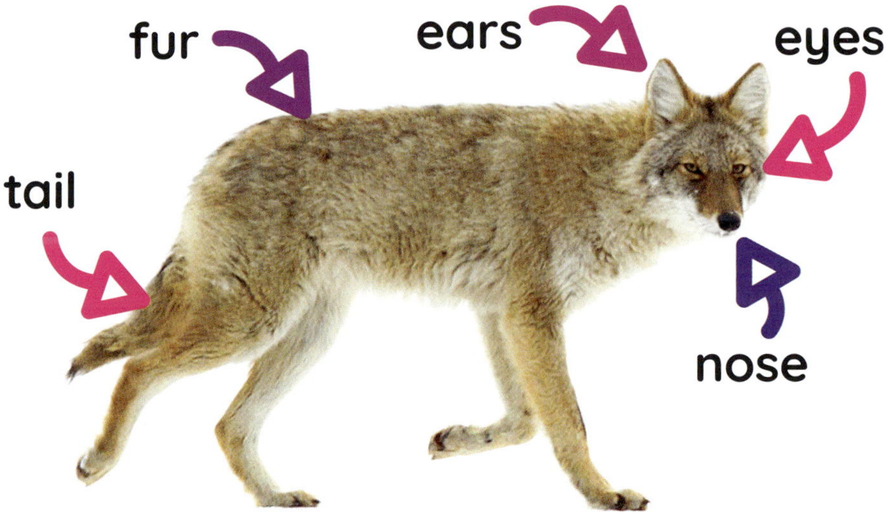

Coyote Food

mice

birds

rabbits

Glossary

dens

homes for animals

prey

animals that are hunted

pup

a baby coyote

To Learn More

ON THE WEB

FACTSURFER

Factsurfer.com gives you a safe, fun way to find more information.

1. Go to www.factsurfer.com.
2. Enter "coyotes" into the search box and click 🔍.
3. Select your book cover to see a list of related content.

Index

The images in this book are reproduced through the courtesy of: Tom Reichner, front cover; Jack Bell Photography, p. 3; Grindstone Media Group, p. 5; Harry Collins Photography, p. 6; Jeannette Katzir Photog, p. 7; Jim Cumming, pp. 9, 22 (top); karl umbriaco, p. 11; Ziga Camernik, p. 12; Matt Knoth, pp. 15, 22 (mice); Kitchin & Hurst/ Age Fotostock, p. 17; Leo Keeler/ Alamy, p. 19; JayPierstorff, p. 21; critterbiz, p. 22 (birds); Steve Shinn/ Alamy, p. 22 (rabbits); A_Lesik, p. 23 (dens); Rudmer Zwerver, p. 23 (prey); Warren Metcalf, p. 23 (pup).